Technical Marketing®

secrets revealed

Strategy and Tactics for Competitive Results

"The Silicon Valley Survival Handbook"

first edition

by
Craig Thomas Ellrod

Publisher:

Technical Marketing, Copyright © 2008 by StrateQuest, LLC. All rights reserved. This publication contains information protected by copyright and trademark laws. No part of this publication may be used, photocopied or reproduced in any manner whatsoever without prior written consent from StrateQuest, LLC. For information, address to StrateQuest, LLC., 2068 Walsh Ave Suite B2, Santa Clara, CA 95050 U.S.A. SAN: 8 5 7 – 6 8 7 4.

ISBN-13: 978-0-9822570-0-5
ISBN-10: 0-9822570-0-7

Blog: http://technicalmarketing.org
Web: http://www.technicalmarketinggroup.com

ISBN 978-0-9822570-0-5

56995

9 780982 257005

trate
Quest™

Dedicated to the advancement of our profession, the enhancement of your skills, and the discovery of something great.

Contents

Technical Marketing

What is Technical Marketing anyway? It sure sounds really snappy. The term originated at the beginning of the explosion of the commercial Internet. Although, after reading through newspaper articles from the 1960's and 70's, and my fathers journals, I can tell you that the job of marketing technical products has been around for decades. Technical Marketing bridges the gap between engineering and sales. In a sentence it can be condensed to, "the art of bringing together significant research on a product and presenting the Delta features in a dramatic way."

As the computer industry approached the turn of the millennium, there continued to exist a gap between engineering and sales at various high tech companies. Engineers were highly talented at building brilliant and powerful computer systems, and sales were equally proficient at relationship management and generating revenue. System's Engineers often times filled that gap but their focus was on revenue and not marketing. There was an increasing need for marketing talent to bridge the gap between sales and engineering, to take the products engineers had built and articulate them in a meaningful way so that sales people and systems engineers could convert them into sales revenue.

> " Technical Marketing bridges the gap between engineering and sales. "

More and more companies in the Silicon Valley began to hire "Technical Marketing Engineers" to bridge the gap between sales and engineering, and to increase sales revenue. Technical Marketing has become a key role in bringing products to market and a key component in competitive strategy in maximizing industry leadership and revenue.

The discovery of Technical Marketing resulted after having been in the Computer Industry for several years, starting out as a programmer, moving into customer support then technical support. Eventually, I became what was then called a Corporate System Engineer for lack of a better term. The SE to the SE's in the field. It was a critical role bridging the gap between engineering and sales. I remember getting a call in Orange County in the late 90's to be flown to the bay area for an interview for a new position. We had talked on the phone about a new type of position, one that never existed before, it was called "Technical Marketing".

The guy that interviewed me didn't fit the mold. He had long hair in a pony-tail, Vans tennis shoes, tattoos and lived on 44th street in Newport Beach. He had actually flown up to Silicon Valley just to interview me. Funny, because I had been living in Huntington Beach at the time and vowed I would never leave the surf. It was at the turn of the millennium so I guess it was a time for change. We had started what became known as a new era for the technical professional, and for me. The boom of the Internet created increased demand for technical people who understood how to market technology in meaningful and dramatic ways, because market execution had to be more precise to match declining budgets.

Product Marketing Engineers and Corporate Systems Engineers became Technical Marketing Engineers. Technical Marketeer's who could work hard, learn fast, pull together a sizzling demonstration and show off a products "Delta" features in a dramatic way suddenly found themselves becoming the rock stars of their employers. Technical people now had a growing and intense new forum to exercise and build their craft. Technical Marketing was born.

Technical Marketing is about presentation of technical products in striking and dramatic ways. Technical Marketing is about competition and strategy and the methods used to win in the marketplace.

Technical Marketing is hard work, and it is a daily struggle to keep going. When I came to valley I was healthy. When I got married later my wife informed me of some odors and smells that I had never been aware of before. I found myself making visits to the doctor for reasons I could not understand. I don't know many guys who start out in Technical Marketing, or who get suckered into technical marketing, that actually stick with it. It is a demanding career. The profession isn't noble like the other arts, the pay is low, and the hours are long. If you genuinely enjoy the work, you will stick with it, and the work will be rewarding to you and of value to others.

The rewards for me are in peeling back the layers of the onion, not knowing what is to be uncovered next, and finding the little nuggets of opportunity that can turn into big development ideas.

Since I have been doing this, there has never been a dull moment, and I truly love what I do. So can you.

Technical Marketing started in the computer industry and has turned out to be very powerful, not only for competitive marketing, but also for the advancement of technology. Technical Marketing is competitive in nature, and incorporates some competitive marketing tactics, yet takes it to another level with more focus and aggressiveness on winning, with specific tools and deliverables to get you there.

You could argue that it feeds the engine of innovation and invention and ultimately improves the quality of products. On this basis, Technical Marketing, and the practices thereof, can be used in any industry, and there are a few that I can think of where it would help.

Intellectual Property
2

The link between Technical Marketing and Intellectual Property is indissoluble. Through research and becoming intimate with your own products you find areas for improvement. By researching the market and competition, you find areas and ideas for competitive advantage and improvement. These ideas become a groundswell for future intellectual property.

Many companies today have an invention disclosure process where you submit your invention or intellectual property discovery. There are typically groups of review teams that determine if the invention disclosure needs to be submitted through the legal department as a patent application, at which time it becomes bona fide intellectual property for the organization.   In order to compete successfully today, this type of process is absolutely crucial. The more efficient this process is, the better it will be for the organization acquiring the intellectual property rights. The first to file, is the first to claim.

In this era Intellectual Property is far more valuable than physical property. It is also less tangible and easier to abscond with. Nevertheless, IP, or Intellectual Property (not the Internet Protocol), is at the core of a Technical Marketeer's being.

The benefits of Intellectual Property abound. IP allows you or your organization to take high ground in the marketplace, and gives you the legal right to do so, thereby allowing you to stake your claim for long periods of time. Hence, you can reap the benefits of the revenue stream for a much longer period of time, without contention or threat of competition. Intellectual Property comes in the form of Patents, Copyrights and Trademarks and is that medium by which you or your organization can defend itself or stake a claim in the market, should a legal battle arise over ownership of a particular technology.

There are also problems identified with Intellectual Property. It is costly to develop and maintain, it can be stolen and is frequently "*borrowed*", it can be misdirected and miscommunicated, it can be forgotten about, it can be left undeveloped and underresourced. Some really great ideas never get developed because of a lack of funding, or a lack of strategic viability. Someone thinks it won't benefit the organization competitively at the end of the day, and there isn't enough money to finance it.

The Technical Marketeer always documents and keeps records of their work and progress. It is common practice to carry a lab notebook around to document meetings, drawings and ideas. Be sure to keep track of Date, Time, and Names of those involved. This information is needed to fill out invention disclosure forms and documents that describe the inventions so that anyone can understand them.

3
Credibility

Credibility is key. Without credibility no-one is going to listen to you, or believe you. I have seen many enter the field of Technical Marketing because it appeared glamorous and sexy and filled their ego and bank account, yet never perform. I have seen many enter the field thinking they could sidestep the responsibilities of a sales engineer where there is accountability to sales numbers. Technical Marketing is much harder, if not as hard as working out in the field, because you are often called on to go out into the field to help sell the product.

Technical Marketeers have to know what they are talking about technically, which means you need to know your stuff. You also have to put the time in, in the lab, and in the field. You have to be book smart, rack smart and customer smart. I have watched guys who have been in technical marketing for years who still can't plug in an Ethernet cable, or figure out the difference between bridging and routing. These are basics, and you need to go much, much deeper than this. Often times, you have to know more than the CTO of your company if you want to do the job right.

Remember, Credibility is your mantra throughout your Technical Marketing career. Make sure you know what you are talking about. When it comes to the computer industry, always document your test network with detailed diagrams showing data flows, links, interfaces and IP (Internet Protocol) Addresses. Always back up your numbers with data that can be reproduced, *by your customers*. Credibility is of utmost importance to a Technical Marketeer. The only way to solidify your credibility is to back up your claims with evidence from testing and documentation.

Competitive Test

You must perform tests to prove and back up your facts. You can hire someone to do it for you, or you can do it yourself. Caveat emptor, as some of the traditional outsourced marketing test outfits will publish anything you tell them to, because you are paying them. I have read many test reports that lack credibility because they were "bought". A competitive test report must be believable, and you can't pull the wool over anyone's eyes. You can't use smoke and mirrors here. I've seen cases where the report lacks merit because the test methodology isn't even published in the report altogether, which brings into question the credibility of 1) the test house that produced it and 2) the vendor that paid for it. The best way to dig out the details is to test it yourself. You need test equipment to do this.

Specific to the computer industry, stopwatches don't exist in this field. The human eye, human hand and mechanical workings of a stopwatch are slower than the electronics being measured, and cannot provide meaningful and accurate performance data. There are legitimate tools available for measuring performance of products in the field of Technical Marketing.

Test equipment is expensive, so you will have to bite the bullet, and buy it, lease it, find a friend who has it or write your own software to do it. There are some traditional network testing favorites. There are many other up and coming network testing equipment vendors that do a much better job, with much better pricing options and much better user interfaces. There are some freebies on the internet, but they don't scale, don't perform well and don't keep track of results for you.

Performing a competitive test on your products as well as other products is important, because it provides you with the evidence you need to back up your claims. Remember that credibility is of utmost importance to a Technical Marketeer. The way to keep your credibility is to back up your claims with evidence from testing and documentation.

Competitive test involves two important components, the device under test (DUT) (or solution under test (SUT)) and the Testing Equipment. Different DUT's require different test networks. The following are examples of the several variations of physical test configurations that can and have been used in network competitive tests.

Competitive test is a critical component of Competitive Analysis, and is used in the Validation Test/Competitive Test phase.

Vendor	Popular for
http://www.ixiacom.com	IxLoad, IxExplorer, IxChariot for L2-L4 testing.
http://www.spirent.com	SmartBits for L2-L4 testing, Avalanche/Reflector for L4-L7 testing

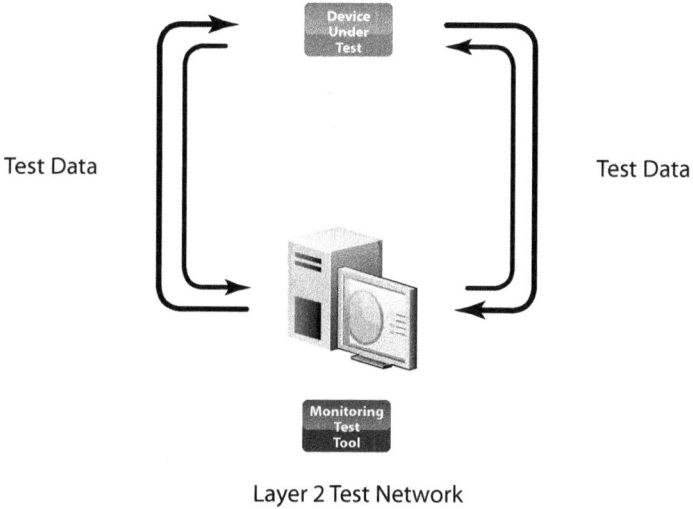

Layer 2 Test Network

TANSTAAFL

"There ain't no such thing as a free lunch". This is especially the case with test equipment. Test equipment is expensive, so you will have to bite the bullet, and buy it, lease it, find a friend who has it or write your own software to do it. There are some traditional network testing favorites listed later on. There are some freebies out there, but they just don't scale and the hard-core IT shops won't buy into the metrics or the test results you produce.

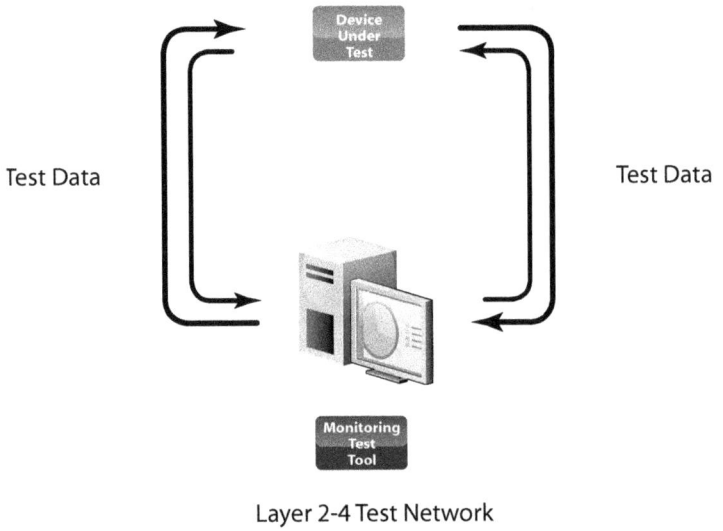

Test Data — Test Data

Device Under Test

Monitoring Test Tool

Layer 2-4 Test Network

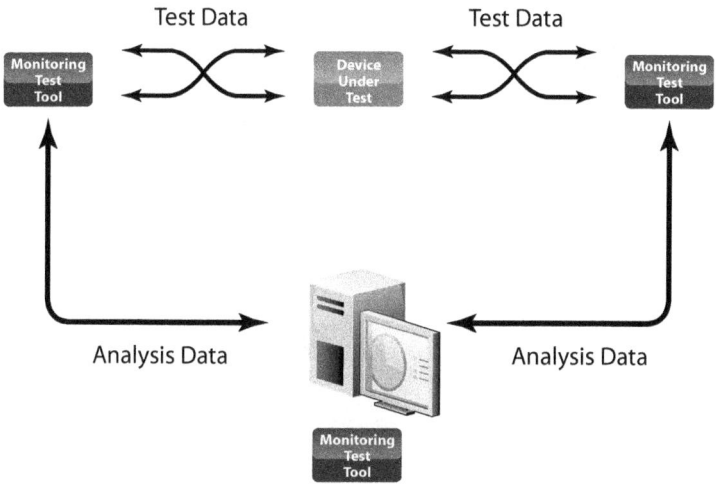

Layer 4-7 Test Network

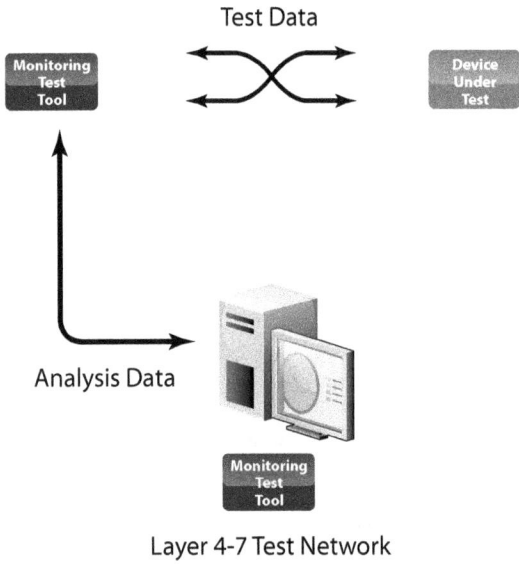

Layer 4-7 Test Network

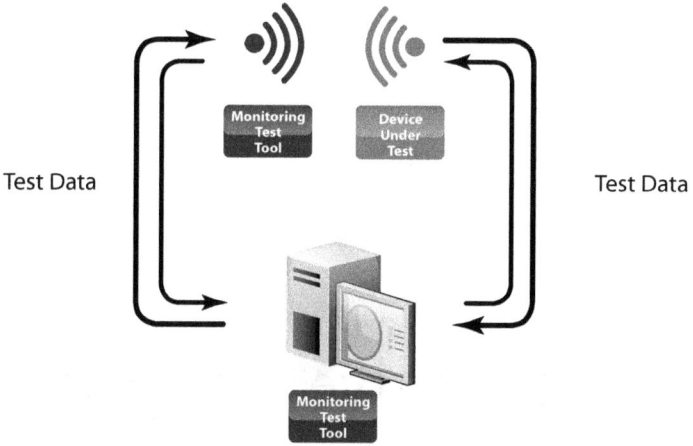

Test Data

Test Data

Layer 2-4 Wireless Test Network

5 Competitive Analysis

Competition is healthy. It is what drives the USA and the global economy's pulse. The government wants you to compete, in fact there are laws in place that protect it. I am not a lawyer so I am not going to expound the law. But I am here to tell you that it is OK to compete, and compete fiercely, as long as you do it legally. If you are not sure of the legality of your position, ask a lawyer.

You were taught in school to use SWOT analysis – strengths, weaknesses, opportunities and threats. Strengths and weaknesses are internal, opportunities and threats are external. This is cute, but not useful for Technical Marketing. It doesn't dig deep enough and doesn't explore widely enough. There is no acronym for digging deep. There is only hard, laborious work.

There are some websites that offer free competitive information. Some that offer competitive information for a fee. You can hire a team to do the competitive analysis for you, in house. Startup companies are sometimes better at competitive analysis than larger companies. Larger companies, for whatever reason, think they are invincible and take a lackadaisical approach to competitive analysis, leaving it in the hands of the sales force to gather what they can while on the run. By the way, having the field gather intelligence is a good, and important tactic for reconnaissance or information gathering on your competition. Partner's and reseller's will often give sales people anything to win business, including information on the competition. I'm not saying this is ethical, I'm just saying it happens.

Once all of the competitive information is compiled, it needs to be disseminated. The means to do this is up to the creative imagination. Some methods involve weekly updates to the field to keep them abreast of the latest competitive information and the location of the repository. Presentations, documents, spreadsheets, graphs and the like are common mediums. The biggest challenge facing competitive analysis is getting the results into the minds of the people that need it. At one company where I worked, we held regular competitive update conference calls and sent out regular messages with competitive briefs and the repository location, and people still called and e-mailed from the field asking for the latest information on *"xyz company"*, when I had given them the information yesterday. Some of the more popular tools coming of age are Wiki's, Blogs and Partner websites, where the information is posted regularly, HTML formatted e-mails are sent out, and links in the e-mail take the reader back to the Wiki's, Blogs or Partner sites containing the latest competitive information.

Competitive Analysis is a lengthy process that can be divided into different phases.

- Research
- Validation
- Analysis
- Presentation

Research

Start with what is publicly available. The internet is your friend and you can gather so much information from it using the search engines and websites of various targets that you are researching.

Tradeshows are designed to generate leads for field sales, but have increasingly become a forum for snoops and lurkers doing reconnaissance on the latest information companies are leaking to the market. There are a lot of imposters at tradeshows sent in to draw out information that would otherwise not be available to the mainstream public. It's an interesting game when someone says they are an independent consultant.

By law public companies have to present financial statements every quarter and at the end of the year. This may or may not help you with the competitive analysis, but can give key insights into the direction a competitor may or may not take based on revenue streams.

Develop your contacts within the industry and go to lunch. Ask questions while you are there. Friends know what is going on with key projects better than any other source. Release dates, hiccups, bugs, roadblocks, key discoveries, important partnerships and mergers all play a part in competitive analysis. It's the little things that you need to pay attention to that can make or break a technology. If there is a key weakness that you discover in a conversation, or a competitive delta that hasn't been exploited, these boondoggles can be well worth the time.

Industry press does the same thing, only they are given key data from companies because they have special privileges, usually because they are going to write a press article and give some type of marketing exposure in exchange. Search these articles for stuff you might have missed, they are juicy.

Although daunting, but also a good source of information, are the various existing patents and trademarks databases online.

Pull all of this information together, take notes and store it in the Public Information File on your computer.

Gather Public Information
From Intenet Search

Gather Public Information
From Tradeshows

Gather Public Information
From Published Financials

Gather Information from
Insiders

Gather Information from
Industry Analysts

Gather Information from
Industry Press

Compile Public Information
Gathered from Research

Research Phase

Validation

Whether you are performing a competitive test or validating your own product against your research, you must perform hands-on lab work. You must be technically competent and go technically deep. There is no way to cheat.

In the same way that you prepare an outline before writing a paper or book, you prepare a lab layout and network diagram before plugging in the equipment. Otherwise, you will end up with a mess. It will enable you to troubleshoot your way around once you have a network diagram.

With regards to the computer industry, the network diagram should include classic elements such as the "Device Under Test" - DUT (or "Solution Under Test" - SUT), and the "Testing Tool". Other elements commonly included in the network diagram are the IP addressing scheme (Internet Protocol), interfaces, speeds, feeds, vlans, trunks, clients, servers, traffic flows, and media types. Most media in use today is Ethernet at varying speeds. A network diagram should include two parts, the physical network layout, and the logical network layout. The physical network layout shows all of the physical connections, interfaces, cables, ports, patch panels, etc. The logical network layout shows all of the higher level connections, such as IP addressing, vlans, trunks, traffic flows, etc.

A clear statement of what is to be tested and the goals of the test should be outlined in the test methodology and test plan, so that someone else can pick up where you left off in case you get hit by a bus. In my case, I often get distracted, and have to refer to my notes.

Building out the physical network takes tried and true know-how. Either that or a desire and strong will to succeed combined with some smarts. Finding the right cables, switches, and other hardware is the easy part. Plugging that stack of equipment in and configuring it along with the DUT or SUT, Test Tools, switches, routers, racks, power supplies and getting the whole thing up and running in record time - is what takes talent. If you can pull this off in a few hours, that which takes the average person a week to do, you might have what it takes.

Once everything is staged, the DUT (or SUT) can

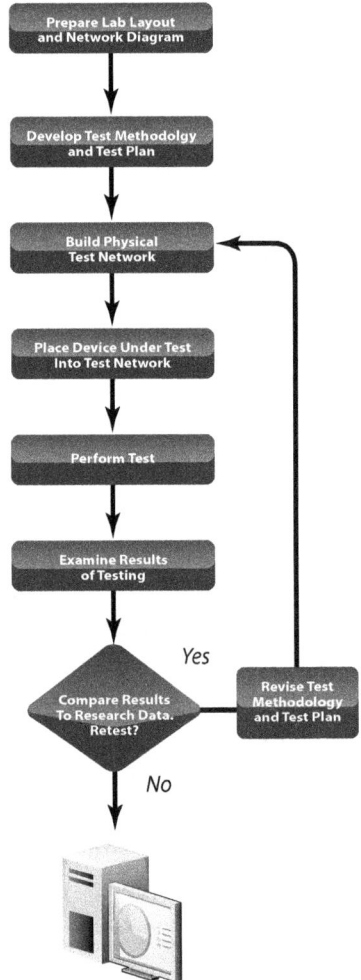

Validation Test Phase

be faithfully placed in-line for test. But, before you do that, I highly recommend you run a test without the DUT just to get a baseline reading. Depending on if you are running "routing" vs. "bridging", this could add extra time and flavor to your test methodology.

Perform your tests, and keep track of them using spreadsheets. Save every test in a separate worksheet or spreadsheet. Every detail counts. Many test tools save results in .csv or spreadsheet format already. Remember, no stopwatches!

After the test run, or several test runs, it is time to examine the data. It is better to have more data, than less, because you might notice a competitive delta that didn't exist before. So pull all of it together and comb through it.

Look for deltas that didn't exist before. Look for high and low variances and things that don't look right or don't add up. If numbers seem to be too askew, you will have to review with a trusted colleague. I always review my results with the engineer that built the product, because they have a gut feel one way or another.

Compare results to research data. If results don't look dramatic enough to "tell a story", then you quite possibly will need to revise the test methodology and test plan (back to the drawing board), and re-test. The point is to dig-up dramatic results; however, honest and verifiable.

Note: Some companies keep their test methodologies, test plans and test results a secret. I think this diminishes credibility. If you have done a good job, and you are the best, publish your test methodology, test plans and test results so that your customers can validate your results. It brings credibility to your organization and your product and will leave your competition in the dust.

Deliverance

The hardest part of Technical Marketing is delivering. You can spend all day in research absorbing information, and this can be fun. You can spend all day in the lab and never get anything done. The key is to develop a plan, sometimes written out in a *"Statement Of Work"*, which outlines what the problem definition is, what it is you need to accomplish in the lab, and what the outputs and deliverables are going to be at the end of the project. In this SOW should be a timeline of estimated workflows, so you can double check your progress and stay on track. Because, just like in show business, you are only as good as your last deliverable. And if you aren't delivering, ...

Industry Analysts

Lets face it, they aren't going to be going away anytime soon, so learn how to use them to your benefit.

Sample Competitive Performance Spreadsheet

L2-4

	Your Organization Your Product Δ
Maximum Concurrent TCP Connections	1,400,000 conns
Maximum TCP Connections/Sec	25,000 cps
Maximum Throughput	1 Gbps
TCP Multiplexing Ratio	10:1

SSL

Maximum Concurrent SSL Connections	100,000 conns
Maximum SSL Throughput	1 Gbps
Maximum SSL Transaction Rate - No Session Re-Use	10,000 tps

L7 HTTP

HTTP Transactions/sec	45,000 tps

L7 HTTPS

HTTPS Transactions/sec	15,000 tps

HTTP Latency

HTTP 1.1 Latency	< 1 ms (TT1B) < 10 ms (per Web Page)

HTTPS Latency

HTTPS Latency	< 10 ms (TT1B) < 1 s (per Web Page)

[1]Some architectures don't match and cannot be validated or tested

[2]Some solutions are so flawed they cannot produce adequate results they are considered "*unproven*"

Competitor 1 Product x	Competitor 2 Product ψ	Competitor 3 Product z
800,000 conns	700,000 conns	600,000 conns
15,000 cps	14,000 cps	13,000 cps
900 Mbps	800 Mbps	300 Mbps
7:1	5:1	not available[1]
100,000 conns	50,000 conns	40,000 conns
500 Mbps	150 Mbps	100 Mbps
5,000 tps	2,500 tps	1,000 tps
30,000 tps	25,000 tps	20,000
10,000 tps	5,000 tps	unproven[2]
< 1 ms (TT1B) < 50 ms (Web Page)	< 1 ms (TT1B) < 50 ms (Web Page)	< 1 ms (TT1B) < 1 s (Web Page)
< 10 ms (TT1B) < 2 s (Web Page	< 10 ms (TT1B) < 3 s (Web Page)	unproven[2]

Analysis

You have all of your research. You've done your hands-on validation test. Now comes the fun part, competitive analysis. Here is where you supplant myth with reality, claimed with proven, not only for your own product(s), but for your competitors as well. Lay it out on the table and see where you stand. This is a good time and place for some quiet reflection in the war room. From this point the light of reality shines on the current standing of your products and future endeavors that you wish to unfold. What can be revealing at times, is that many organizations don't take the time to dig deep and accurately find out where they stand competitively.

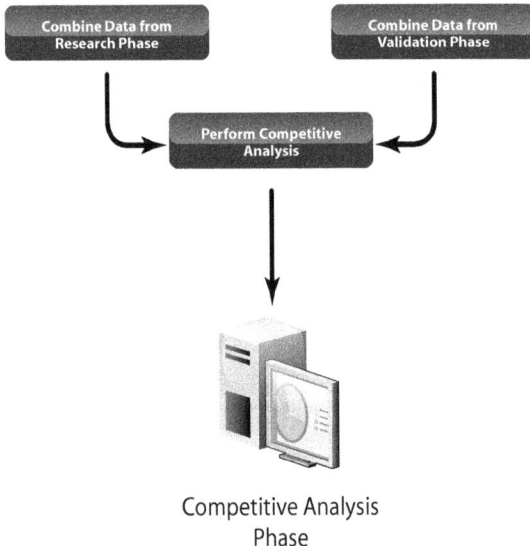

Competitive Analysis
Phase

FUD

Technical Marketeers never throw FUD. Fear, Uncertainty and Doubt are for the less noble and less adept. Everything we write and talk about has to be backed up by facts. This is your job as a Technical Marketeer, to bring credence and credibility to the technical marketing data at hand. Whether someone else has done the fact checking or testing, or you spend hours verifying the facts yourself in a lab, no FUD. Hence, your bullet is sure to pierce when it hits the target.

Claimed

Theoretical metrics based on engineering specifications.

Proven

Real world testing results. Metrics obtained through real world testing.

Margin 85

Margin generates Profit. Profit is directly tied to stock price, if your company is public. It is equally important if your company is private. How do you set the selling price of your product? For some companies, the rule of thumb is 85% margin over and above the Cost of Goods Sold. Software enjoys a 99% ~ 100% margin in some cases because there is no hardware manufacturing cost. Older products in a mature market generate lower margins.

Presentation

The results of the analysis can be presented in many forms. Credibility is key, the presentation of data cannot contain any FUD or invalid data, or the field will lose trust and credibility - hence they will lose. FUD and invalid data is a weakness. Make sure your data sizzles, but make sure it is accurate.

As mentioned before, Technical Marketing involves presenting your data in "Dramatic" ways. This means not only knowing the technical aspects of the product, but knowing how to present it. There are many exciting mediums to make use of today. The simplest of which most people use on a daily basis - Microsoft Office. A competitive brief, technical document or white paper can be typed up and printed to Adobe PDF format. Granted, it gets messy when sophisticated graphics are required, but it gets the job done.

Competitive Playbook

Much like you see in the hands of an NFL coach on the field during a game. A playbook lists all of your products, offensive plays, defensive plays and counter attacks to competition. Knowing your competition is important, but knowing that your competition knows you and your plays is more important. You need to know how to counter attack their attacks, defenses and trump them before you get to the playing field.

Powerpoint is good for presentations. There are many programs for the delivery of video based content and demonstrations - the crux of a Technical Marketing presentation. Technical

Marketing is all about giving demonstrations in a dramatic way, which is why screen recorders such as Camtasia and uTIPu are so popular. You can build a dramatic demonstration offline, and let it run indefinitely on a Blog or Tradeshow floor. With Virtual Tradeshows becoming more popular, Virtual Demonstrations will be a common venue for Technical Marketeers.

When it comes to making presentations dramatic and professional, there is no substitute for the Adobe tools, and the common format for distribution is the Portable Document Format (PDF). All Adobe and Microsoft products output to PDF. Most all computer systems can read PDF.

Some of your material is good for public consumption, while some is even more important to the internal sales team. Nothing is more important than the competitive playbook.

Much like you see in the hands of an NFL coach on the field during a game, the playbook lists all of your products, offensive plays, defensive plays and counter attacks to the competition. Knowing your competition is important, but knowing that your competition knows you and your plays is more important. You need to know how to counter attack their attacks, defenses and trump them before you get to the playing field. A highly researched, tight, focused playbook is crucial to field sales. Most playbooks only contain the top 3~4 competitors, leaving the others for the lengthy competitive briefs and competitive presentations.

Just as valuable are regular updates in the form of competitive briefs, or announcements to the field. They should be short, one to two pages that highlight the competitive landscape and weaknesses of the competitor.

Competitive Analysis often finds it's way into intellectual property and strategic business development. It is also useful in competitive presentations, RFI's, RFQ's, RFP's, responses to analyst reports, press releases, white papers, training material, sales presentations, industry reports, Market Requirement Documents (MRD), Product Requirement Documents (PRD), product roadmaps, partner enablement, and of course performance analysis documents.

Data from competitive analysis can also find it's way into data sheets, frequently asked questions (FAQ) documents, hot sheets, battlecards, fast fact sheets and customer presentations. The practical usefulness of the competitive research serves many needs, but the two that keep recurring are fighting the everyday battles in field sales against the competition, and long term strategic planning. If you don't have the facts, you won't be successful at either.

Competitive analysis data has been finding its way into online knowledge bases and reports for free or fee. The most compelling and useful competitive analysis documents that make use of this data are the competitive briefs and the competitive playbook. Playbooks should be short and concise, because on gameday, your troops will have already done their month long studying of the competition, they just need factual bullet points as reminders for execution. If it is too long, people won't read it. Getting to the customer first is the best strategy, because you can play your offense, and "*lay your traps*" for the competition, leaving them tongue tied and unprepared when they do arrive.

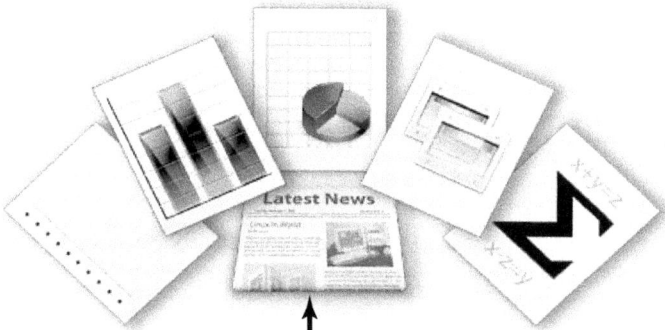

Presentation of
Competitive Data

Sample Statement Of Work

Project Overview

A brief and overview of the project

Project Objective

The Objective is to:
- Show the value of x and y solution working together
- Asses Scalability of x and y solutions
- Test and prove performance of x and y solutions

Project Owners

Stakeholders and contributors who are accountable for the deliverables at the end of the project.
- Name, e-mail, phone.

Project Environment

Details for the lab environment:
- Physical environment (IP Addresses, etc)
- Logical environment
- Diagrams
- Detailed list of products and versions

Project Deliverables

The output at the end of the project.
- Interoperability report, Scalability report, Performance report
- Competitive Brief, Competitive Playbook
- Video tip, Demonstration script, White Paper

Project Timeline

Estimate of project workflows and workplans. Use a spreadsheet.

Estimates, Expenses and Change

- List Responsible parties for costs and expenses.
- Provide a change management process, to CYA.

Engagement Agreement

- Stakeholders and Contributors Sign or Agree on the SOW.
- An agreement is in place to keep the project goals on track.

Sample White Paper

Introduction

A brief Introduction

Problem Statement

Describe the problem you are solving
- List all of the problems you are solving.
- Customers have pain, so address it here.
- ...

Previous Options

Discuss how it used to be
- The risks and downsides to staying the course and not changing.

Proposed Solution

Solutions have to provide Benefits, not features, so write about them. Talk about how you solve the problem.
- Benefit 1
- Benefit 2
- Benefit 3

Summary

Summarize
- Restate the problem
- State how you solved the problem

Conclusions

Be brief, be concise, be a thought leader.

Sample Invention Disclosure

Title of Invention

Title is usually descriptive of the invention.

Inventors

Name, Address, Citizenship, Department, Contact Information.

Background of the Invention

Describe in general terms the purposes and object of the invention.
- Include drawings, flowcharts, sketches, note pads, papers, meeting minutes, meeting notes, diagrams, etc., which help with understanding and evaluating the invention.
- List Prior Art or Closest known technology.

Description of the Invention

Describe the Invention in detail so that another competent person in the field would be able to understand it.
- Describe each step in the process and what it accomplishes.
- Provide test results, List any referenced patents or publications.

Advantages of Invention

- How Invention differs from Prior Art.
- Describe the problem that the Invention solves.
- State what is "New" in the Invention.
- List the advantages of the Invention over the current state of the art.

Alternatives

- Describe other forms of the invention.
- Other means of accomplishing the same thing.

Record of Invention

Date first conceived?, Physical record? Whom did you disclose to? Disclosure date?, Evidence of disclosure?

Public Disclosure

- List papers, abstracts, internet postings, presentations, sales, offers to sell, planned readiness, etc.
- Public disclosure sometimes results in loss of patent rights.

Signature

- Name, Signature of Inventor, Date

Sample Competitive Brief

Executive Summary

A brief and concise description of the findings.
* The three top bullet points that readers should be aware of
* ...
* ...

Discussion Points

A list of all key findings from competitive analysis in paragraph or bullet form.
* List all of the weaknesses
* Some readers need to know all of the dirt
* You may need to refer back to this during a heated, drawn out competitive battle.
* ...

Where does the competitor mislead the customer

List the juggernauts
* They always leave something out, that could cause the customer some grief
* ...

Performance

Every customer deserves to know the true performance
* Educate the customer on what the competitor hasn't told them
* Cite your sources, even if you did the testing
* ...

Pricing

Its always about money
* Compare the price of the solutions in column format side-by-side
* ...

Conclusions

Be brief, be concise, be fierce.

Sample Competitive Playbook

Offense

Propose to customer	Your Company, Your Product Delivers
Key Benefit • Concise description Why? • Explain the compelling Impact to customers business	• Delta Feature Δ^1 • Delta Feature Δ^2 • Delta Feature Δ^3 • ... • *etc*
...	...

Defense

Customer will ask you
Key Benefit • The competition has ...
...

Competitor 1
Response,
Your Counter Response

Competitor 2
Response,
Your Counter Response

Competitor 3
Response,
Your Counter Response

Competitor:
- Our Product does x^1
- Our Product does x^2

Your Response:
- Reveal facts that disprove the claim
- Remind customer of weakness of competitor
- Refocus customer on your Delta features
- *etc*

...

Competitor:
- Our Product does ψ^1
- Our Product does ψ^2

Your Response:
- Reveal facts that disprove the claim
- Remind customer of weakness of competitor
- Refocus customer on your Delta features
- *etc*

...

Competitor:
- Our Product does z^1
- Our Product does z^2

Your Response:
- Reveal facts that disprove the claim
- Remind customer of weakness of competitor
- Refocus customer on your Delta features
- *etc*

...

Competitor 1 Claim,
Your Counter Response

Competitor 2 Claim,
Your Counter Response

Competitor 3 Claim,
Your Counter Response

Competitor:
- Our Product does x^1
- Our Product does x^2

Your Response:
- Your product does Δ^1, Δ^2, Δ^3
- Call out your Delta Features
- Remind customer of weaknesses of competitor
-
- *etc*

...

Competitor:
- Our Product does ψ^1
- Our Product does ψ^2

Your Response:
- Reveal facts that disprove the claim
- Refocus customer on your Delta features
- Remind customer of weaknesses of competitor
- *etc*

...

Competitor:
- Our Product does z^1
- Our Product does z^2

Your Response:
- Reveal facts that disprove the claim
- Refocus customer on your Delta features
- Remind customer of weaknesses of competitor
- *etc*

...

Sample Organization Profile

Organization/Company

Name

Address

Contact

Public or Private

Founded

Headquartered

Number of Employees · *(geographic split, departmental split)*

Annual Revenues · *(last three years. profitable/unprofitable)*

Expected Revenue Growth · *(for next twelve months)*

Rounds of Venture Capital · *(include all rounds of VC raised)*

Debt

Cash

Profit Ratio on Product(s) · *(ex: 85% profit margin)*

Mission Statement

Product

Current Product(s)

Version(s)

Release Date(s)

Software, Hardware, Both

Feature Set

Architecture

Integration with Standards

Availability and Performance

Sample Organization Profile

Product...cont'd

Target Market

Vertical, Horizontal, Both

Market Driver   *(key market opportunity)*

Value Proposition   *(why the customer buys)*

Planned Features   *(next set of advanced features)*

Missing Features   *(according to your customers)*

Customers

Number of Customers   *(by geography)*

Number of Units Sold

Reference Customers   *(list the top five)*

Key Delta Features   *(ones that solve customers problems)*

Typical Customers Profile   *Large Enterprise (over1000 users)*
                                 Small to Medium Business (under 1000 users)
                                 Service Providers

Typical Customer Sale

Customer Satisfaction Rating

Strategy

Sales Strategy   *(direct, channel, partners, combination)*

Key Partners

Key Competitors

Pricing of Product

Annual Licensing Cost

Annual Support & Maint Cost

6

Delta Features

Delta features take you to high ground. They fall out of the competitive analysis. Delta features combined with competitive analysis form the basis for Technical Marketing. Find features that your competitors don't have, and use them to your competitive advantage.

Some companies just outright buy smaller companies to acquire Delta features to gain high ground in the marketplace, also known as growth through acquisition. Getting bought is an exit strategy that many smaller companies seek, once they have fought well in the marketplace.

Delta features are those that competitors don't have, and you have a confirmed need for them from your customers. They can involve everything from physical design, usability, ease-of-use, and practical applicability to performance.

A good example of this is when the traditional router company in the valley offered it's operating system on a single monolithic software image, which when problems occurred would crash and take down the entire router along with all of the connections on that router.

Switch competitors developed the first modular operating system, with components linking into the switching and routing operating system that if a problem occurred, the independent modular systems would remain running while the problematic module would be troubleshot. This was a huge Delta and allowed competitors in the Silicon Valley to shift a significant portion of market share from the incumbent.

Marketing Warfare

Technical Marketing is War. Dig up the dirt on your opponents and exploit them to no end. Play to win every battle, and conquer the war. However, there are times when you need to recognize when you can't win, and live to fight another day.

Marketing warfare is like a full continental confrontation intertwined with a series of battles. Some short, some long. This isn't new information, and it probably won't change anytime soon, because we are all trying to win by hoarding the most money at the end of the game. When is the end of the game?

You have to pick your battles, but as a Technical Marketeer your job is to help win the war by winning the battles. There are many theories and strategies to fighting these battles. And there are a lot of books to read to add to your arsenal. I try to keep my list short and simple and use that which I know works, because most of my time is spent in the trenches fighting and strategizing. Some of my favorites for inspiration over the years have been:

- The Art of War
- The Art of Peace
- Marketing Warfare
- Blue Ocean Strategy

There are only a couple of large companies in Silicon Valley that are perverse enough to use "The Art of War" against it's own employees to tear itself apart from the inside out. If you are going to use any of these tactics, don't use them on your fellow employees, use them on your competitors. Recommending The Art of War to a competitor is a good competitive tactic, because it has the potential to send them on a course for self destruction. Just be sure that you understand "The Art of Peace" and when to use both. There are some good lessons in The Art of War, however, there are better lessons in The Art of Peace.

I remember inventing the phrase in college, "*Quiet Poison*". When I think of Technical Marketing, I always think of that phrase. There are always going to be a handful of well known strategies. It is up to you to find some not so well known strategies that work for you, and keep them to yourself as "*Quiet Poison*" to help you win your battles. Remember, you are fighting your competitor, not your brother in arms.

Some of the older, well known strategies from The Art of War, "Gain the advantage and you win, lose the advantage and you die", "Complete victory is when the army does not fight", "Those who know when to fight and when not to fight are victorious", "A victorious army wins first and then seeks battle", "Induce opponents to come to you, and their force is empty", "Incite opponents to action in order to find out their patterns of movement and rest", "Foreknowledge must be obtained from people who know the conditions of the enemy", .. and probably the best of them "Keep

Developing Strategies

When you and your competitor have both read all of the strategies together and are up on the game, it's time for you to discover a new strategy, or look for another unsuspecting one. Because they are going to be using the tactic that you just read about, against you. When you find a method or tactic that works against the competition, you keep it close.

your friends close, keep your enemies closer".

One of the best lessons that can be learned from "Marketing Warfare" is that strategy should be developed from the ground up, not the top down.

In playing offense and defense, you will get attacked and will need to counterattack, that is given. Your competitors

The Best Marketing Tactic

The best marketing program that is in use today, and it continues to kill the big gorillas, is 'its free'. Give your software away, proliferate the market with it. Make it abundant, useful, and credible. If the market accepts it and it is good, it will multiply. Now you can use your imagination on how to draw revenue from this model and companies are doing it successfully. Large companies cannot compete because they need big margin's to support their infrastructure.

are anticipating counterattacking you. It used to be that smaller companies attacked and larger companies defended. This is no longer the case. Neither size nor market share no longer dictate whether to play offense or defense. These tactics can be used in offense or defense to achieve the overall strategy.

The War Room

Set up a war room. If you are serious about competitive marketing and want to do it right, you will have a war room, with access limited to only those on the team. Think it's silly? This is how WWII was fought and won. Get whiteboards installed on all the walls from ceiling to floor. Map out your plans, put up the timelines. Use large 3'x4' post-it boards that can be pasted to the wall, torn down and saved in case they need to be re-used in a patent application.

G2

"Competitive Intelligence" - The term G2 has been around for many years. G2 is the name of the intelligence branch of the military. In modern tech marketing speak, Getting the "G2" on the competition refers to getting the inside information on a competitor that could help your team win the battle, or even the war. How? If you know what your competition has secretly planned to come to market with to strategically "leap-frog" your team, you can "go-beyond" them even further or secretly take high ground without them suspecting it. Thus, G2 has evolved from the wartime battlefield to the modern day marketing battlefield, and is commonly used to refer to *"Competitive Intelligence"*.

Strategy	Tactic
First Call *"Rooster Call"*	Get to the customer first, at the break of dawn, others have to follow in the wake of your tail.
Frontal	Find the weakness in the competitors strength and attack it.
Flank	A move into uncontested ground, you've found a feature or new market opportunity and make a move for it.
Fence	Find a market small enough to defend.
Block	You missed an opportunity, but recover by copying the competitors move.
Neutralize	Make your competitor irrelevant.
Take High Ground	Take high ground and you win.
Attack Yourself	Improve your position by obsoleting existing ones, re-invent yourself and your products.
Value not Price	Focus on Value, not Price.
C-Level	Establish relationships with top level executives.
"Quiet Poison"	The one that everyone else missed.
Retreat	Know when to leave the battlefield.

Effect

Getting to the customer ~or~ market first allows you to lay your traps and establish your "*mantra*" as industry standard. Competitors struggle to "*match*" the bar you have set.

Attacking their weaknesses won't take them down, attacking their strengths will. All competitors have weaknesses in their strong points. Attacking their weaknesses doesn't hurt either.

Provides an element of surprise, catching the competition off-gaurd, leaving them tongue-tied and spinning precious cycles to catch up to you. Through relentless pursuit, a new market can achieve critical mass.

Reduces the size of the battleground, so the larger competitors can't attack you.

Allows you to maintain your position and keep competitor from getting established. When you are the leader, it is easier to do this.

Remove credibility of the value of the competitor and/or their product.

High Ground is being #1 in mind-share and market-share. It is easier to defend high ground, so you can focus energy on other strategies.

It is harder to hit a moving target, competitors struggle to keep up.

Don't get into a price battle, you will lose - "*money*".

Establish yourself as an industry thought leader and trusted advisor, and it will be easier to get meetings and get them to sign your purchase orders.

A competitor killer, you keep to yourself and within the team.

Save your resources to live and fight another day.

8 Creating Value

Your job as a technical marketeer is to find value, exploit it and market it endlessly. Your job is to find the key "Deltas", or value points that your product has over the competition, and exploit those "Deltas" and market them.

Caveat; I've seen many companies slapped together like art projects overnight, run through some investor funding and never really grab acceptance in the marketplace. The product or service you sell must be tangible and it must be fundamentally good. One of the startups I worked for in the valley had a product that was fundamentally good, in fact excellent compared to the competition, and it just needed to be exploited and marketed. It was pure joy pulling the competitive "Deltas" out of that product, marketing them, and slamming the competition. We really put the heat on the competition.

Part of the art of being a technical marketeer, is finding what exists that is already valuable and exploit it. If no value exists, it is your job to articulate the value to those who can take action to create it. If it is your job to create value, then you are in the wrong position and need to move into product management. By the way, the move from technical marketing to product management is a very natural step and happens quite often in the business. Once an individual knows the details of a company's product inside and out, they are then in a tremendous position to leverage that into creating the next generation of knockout products. Not all companies realize this effect, yet the payoff is big. All too often companies hire the old tried and true product manager that is better at managing schedules, spreadsheets, vendors and meetings, than at creating vision and inspiring brilliance among the product teams.

Part of the function of competitive analysis is to show you where your product stands relative to the competition. As you gather data, and validate through research and testing, you will uncover weaknesses in your own product(s). This information is not to be let out to the public, although your competitors will find out about it if they haven't already, and will use it against you.

When you discover a weakness, it becomes an opportunity for improvement. Not only an opportunity to improve, but to leap frog the competition. When a you find that you have a weakness through your analysis, you prepare the same competitive brief that you would prepare to the field, however, this brief is only shown to product managers and top level executives so that they can take appropriate action. Mark the brief confidential and not for distribution.

Value is in the eye of the customer. You cannot create value if the customer does not see it your way. So, while creating value, it is of utmost importance to validate your Delta's with a real need, that is, customers who will buy it. The best way to do this is to get in front of the customer, present the ideas and get feedback. Feedback can be obtained directly from field sales, system engineering and customer support who know the direct pulse of the customer base. The delicate balance is gathering feedback without revealing too much forward thinking intellectual property such that competitors might get hold of it. There is otherwise no substitute for face-to-face meetings with customers.

Maximizing Spin

My good friend from Montreal who I used to travel with frequently, would always say to me, "It is in the *speen*"! What he meant was, it is in the "Spin". The key to flavoring your sell. You have heard of spin doctors and spin masters, or spin meisters. Well, prepare to become one. I am at a disadvantage having only learned one language throughout my career; however, I know it well. Take the time to know every nuance of your language as you will need to speak, write, type, scribe, record, compose, create, letter, phrase and voice your thoughts, actions and words in the most prolific and passionate manner possible.

Remember the interview question right out of college, "sell me this pencil"? I'm not so sure what kind of answer the interviewer would expect. The answer I think they were looking for was not so much a description of the pencil, but a demonstration that you had tapped into your deep creative inner being, dug deep down in your heart, felt the passion within, pulled up every sizzling, enticing word from the depths of zeal, and exhaled a sweet sentence in an iambic pentametric song in which, when finished, that person had been mesmerized into following you home. I know that's a little over the top, but you get the picture.

Maximizing spin is about passion of syntax, words, delivery and intonation. It is equally opposed to melancholy. It is about selling the sizzle. But what gives your language spin, is the passion that comes from within. When you have the ability to match a prolific writing and speaking style with your passion, you can become a spin master.

The language of business is English, so learn it well. I have worked with people from Montreal, Brazil, Africa, Croatia, Russia, Singapore, Japan, Australia, the Middle East and India, and they all speak English. I have travelled to Germany, France, Ireland, Spain, Sweden, China, Thailand, Mexico, Canada and of course England for business and the people there all speak English. Even on the Great Wall of China, they speak English.

Technical Evangelists

Evangelism has it's roots in religion and the promoting thereof. Technical Marketeers are Evangelists, being those who enthusiastically promote technology in an attempt to build adoption in a market space.

Native languages and dialects are wonderful and interesting and make the world a wonderful place to live and visit, but when it comes time to do business, everyone speaks English. It is important to master the English language in addition to the technical skills of racking and stacking equipment and making it come together as a working solution.

Arguably one of the best marketing documents ever written in the history of the world has it's roots in Latin, as does the English language. Becoming a Technical Evangelist and evangelizing your word is the great aim of technical marketeers.

A dramatic writing style comes with practice, there is no way to cheat. There are tools to help. I use a dictionary, a thesaurus and the web constantly for looking up definitions to inspire my thoughts when writing. Plagiarism is not allowed, your pen and tongue must be unique. Spell check and thesaurus are built into the Adobe and Microsoft products.

Mastering the spoken word as well as the written word is part of the delivery of spin. Writing and creating art with passion is only half of it. Speaking is the other half. I still have a fear of audiences, but not as much as I used too. I used to get tongue tied, sweat profusely and not make any sense. You have to get past that point, so that you know what you are talking about and speak with enthusiasm and power, so that people at least think you know what you are talking about.

Practice, practice, practice. Even the most polished executives continue to practice. I remember getting started with Toastmasters. I've taken public speaking classes and I recommend you do as well. In fact, one of the best recommendations given me was that I enroll in acting and improvisation classes, which I did. Not only was this fun, but it brought out a funny side of me, and allowed me to be comfortable in my own skin on a stage, or video camera in front of an audience.

10 The New Marketing

I'm an old-school Technical Marketing Engineer, and one day I read a job posting for "Technical Marketing" which alluded to someone who knew how to use Adobe Photoshop and could design Web Pages. I laughed at the description while a lady friend of mine did not. I took that moment to heart, and realized that technical marketing may be turning a corner. The skill sets were at different ends of the spectrum. I went back to school to learn the Adobe products. Lets face it, Microsoft word and publisher don't hold a candle to the Adobe tools. They don't anchor and render images and text properly. Adobe doesn't support Microsoft, and Microsoft doesn't support Adobe, and probably never will. Your career depends upon the quality of your work. Once I learned the Adobe products, I kept looking out across my horizon to keep it from becoming a bore-izon. Technical Marketeers must always be on the lookout for creative ways to evolve our art and craft.

Technical Marketeers have to be positive leaders for change in their field, because many times you are being looked at to take that role. As my father always used to say "Lead, follow or get out of the way", being a retired industry executive.

Web $\mathcal{X}$.0 is all about fresh, fast and exciting content. This is a medium we can take advantage of with greater ease and efficiency than ever before, because you don't have to learn HTML, although it helps. The opportunities and venues for technical marketing are turning a corner. There are new places to market ourselves, our products and services. They are coming in the form of community websites, social websites, viral marketing, acceptance based marketing and value based marketing which carries with it a high credibility rating factor.

By now everyone already knows HTML and CSS. The cost of hosting is so inexpensive, you can host many sites and domains on one service provider for pennies on the US Dollar. This is leverage.

Many startup companies and many investment funds are now outsourcing what was thought to be a talent that was not outsource-able – Marketing. I think you will see a lot more outsourced technical marketing, outsourced product marketing and outsourced product management to professional firms that can hire talented individuals that can handle 4 to 5 projects per person, at a lower cost to the company paying for the outsourcing. The talented individual can actually make a higher income than if they were hired full-time for one company. I know that I could personally handle 4 to 5 technical marketing projects for 4 to 5 companies, which is a far better use of my time, and a better use of the marketing expense dollars for those companies, not to mention the reduction in employee expense on the books.

New Tools Being Used

Wikis	Dynamic websites that power community websites, capable of accepting fresh, richly formatted content. Also used for Knowledge Management systems.
Blogs	Short for Web Log, is a dynamic web site, with commentary, events, descriptions, graphics, videos in a continuous chronological order. Blogs contain videos, photos, music, podcasts... everything 'new' related to marketing. Blogs are the fastest way to get your word evangelized and indexed into search engines.
RSS Feeds	Really Simple Syndication. Web Feed formats used to publish blogs a related works around the web quickly as they happen. Used for blogs, news headlines, audio, video.
Forums	Internet forums, or message boards are online discussion sites that evolved from the old bulletin board systems. They are now web applications where you can spread your word with rich content, and get it indexed into search engines. Forums are usually part of community websites. Most large companies now have community websites.
Widgets	A small chunk of code that is embedded in a web page or on a computer desktop. Now these are being used for marketing or as tiny little portals to draw people to main websites, especially for revenue generation.
Gadgets	Cool and dynamic content for websites and desktops, similar to widgets, and serve the same marketing function.
Video Tips	Short videos that can be embedded within websites, blogs, forums, wikis for demonstrations, evangelism, news headlines, etc.

New Tools Being Used

Voice Overs	Put some voice on top of the video tip, and it doesn't have to be your original, there are tools to obfuscate. Make yourself sound like a sportscaster.
Podcasts	Audio or video distributed over the Internet by RSS Feed to portable media players and personal computers.
Webcasts	Audio or video distributed over the Internet using streaming media technology. Much like a radio broadcast, a Webcast can be distributed live or recorded.
Virtual Tradeshows	Replicate the activity and impact of the real thing, using the Internet, driving down costs, saving time for attendees, and zeroing in on targets for prospectors.
SEO (rank juice)	Search Engine Optimization - The art of getting listed at the top of page when someone uses a search engine. The key to getting to the customer first, the best strategy to winning.
Social Networks	Networks with values, visions, ideas, dependent upon the people that tie them together, is now a concept taking shape as websites and forums as a venue for marketeers.
Community Websites	Community websites are popping up all over the Internet to bring together groups of similar beliefs, resources, preferences, needs, thoughts, learning for the greater good. They are turning out to be places for credibility proving and acceptance based marketing. Customers now have a forum to publicly voice approval, disapproval and help others. This is powerful for those who can harness it.
On-Demand Publishing	Choose from several, do a search online. Writing a book, producing a DVD or Video can be done on your own, quickly, inexpensively. You don't need an advertising firm to do marketing.

11

Tools of the Trade

Vendor	Tool

L2-L4 Testing

Vendor	Tool
IXIA	IxLoad, IxExplorer, IxChariot
Spirent	SmartBits

Network Monitoring

Vendor	Tool
Wireshark (Ethereal)	Wireshark
Wildpackets	OmniPeek
Microsoft	Network Monitor
Network Instruments	Observer
TCPDump	tcpdump/libpcap
Etherape	Etherape
Solarwinds	Solarwinds

Web Application Testing

Vendor	Tool
Spirent	Avalanche & Reflector
Load Runner	Loadrunner
WAPT	Web Application Testing Tool
Microsoft	Web Application Stress Tool
cURL	cURL
Shunra	Shunra Website Tester
Microsoft	Tinyget IIS 6.0 Resource Toolkit

Web Application Security Testing

Vendor	Tool
Cenzic	Hailstorm
HP (SPI Dynamics)	WebInspect
IBM	AppScan
MileScan (Paros)	Web Security Auditor

Web Applications with Built-in Security Vulnerabilities

Vendor	Tool
OWASP	WebGoat
McAfee (Foundstone)	HacmeBank & Others
Badstore	Badstore

Location

http://www.ixiacom.com

http://www.spirent.com

http://www.wireshark.org

http://www.wildpackets.com

http://www.download.microsoft.com

http://www.networkinstruments.com

http://www.tcpdump.org

http://etherape.sourceforge.net

http://www.solarwinds.com

http://spirent.com

http://HP.com

http://www.loadtestingtool.com

http://www.download.microsoft.com

http://curl.haxx.se/

http://www.shunra.com/website_testing.aspx

http://download.microsoft.com

http://www.cenzic.com

http://www.hp.com

http://www.ibm.com/developerworks/downloads/r/appscan

http://www.milescan.com

http://www.owasp.org/index.php/Category:OWASP_WebGoat_Project

http://www.foundstone.com/us/resources-free-tools.asp

http://www.badstore.net

Vendor	Tool
Web Application Monitoring	
OWASP	WebScarab
Paros	Paros Proxy
Mozilla	Live HTTP Headers
Fiddler	Fiddler
Microsoft	WFetch
IE Inspector	IE Inspector
IE Watch	IE Watch
Wireshark	Wireshark
Web Services (SOA) Testing	
MileScan	Web Service Analyzer
IBM	AppScan
Parasoft	SOATest
cURL	cURL
WAN Simulation	
Apposite Technologies	Linktropy
Shunra	Shunra
VoIP Testing	
Shunra	Shunra VoIP Testing
Spirent	Spirent VoIP
IXIA	IxVoice
Network Instruments	Observer
Wireless Testing	
Veriwave	Veriwave
Spirent	Spirent Wireless & Mobile
Shunra	Shunra Wireless Testing
Lightpoint	Lightpoint
Network Instruments	Observer

Location

http://www.owasp.org/index.php/Category:OWASP_WebScarab_Project

http://www.parosproxy.org

http://www.livehttpheaders.mozdev.org

http://www.fiddler.com

http://www.download.microsoft.com

http://www.ieinspector.com

http://www.iewatch.com

http://www.wireshark.org

http://www.milescan.com

http://www.ibm.com/developerworks/downloads/r/appscan

http://www.parasoft.com/jsp/products/home.jsp?product=SOAP

http://curl.haxx.se/

http://www.apposite-tech.com

http://www.shunra.com/network_emulation.aspx

http://www.shunra.com/voip_testing.aspx

http://www.spirent.com

http://www.ixiacom.com/products/ixvoice

http://www.networkinstruments.com/

http://www.veriwave.com

http://www.spirent.com

http://www.shunra.com/wireless_testing.aspx

http://www.litepoint.com/

http://www.networkinstruments.com/products/observer/wireless.html

Vendor	**Tool**
Security Testing	
IXIA	IxANVL, IxVPN, IxLoad, IxNetwork
Spirent	Spirent ThreatEx
NMAP	Network Mapper
Nessus	Nessus
Metasploit	Metasploit
HPing	HPing
Kismet	Kismet Wireless
DSniff	DSniff
...more at...	
Network Tools	
Solarwinds	Subnet Calculator, TFTP Server
Wildpackets	Subnet Calculator
NetInfo	IP Scanner
Colasoft	MAC Scanner, Ping Tool
Radmin	IP, Port, LAN Scanner, IP Calc
IpSwitch	Ping Pro Pack
IpSwitch	FTP Client & Server
FileZilla	FTP Client & Server
SmartFTP	FTP Client
Cute FTP	FTP Client
Core FTP	SFTP, FTP
WinSCP	SFTP, FTP, SCP
Firefox FireFTP	SFTP, FTP
PuTTY	SSH & Telnet
Open SSH	SSH
Bitvise	Bitvise Tunnelier SSH Client
Bitvise	Bitvise WinSSHD SSH Server

Location

http://www.ixiacom.com/solutions/testing_security

http://www.spirent.com

http://nmap.org

http://www.nessus.org/nessus

http://www.metasploit.com

http://www.hping.org

http://www.kismetwireless.net

http://www.monkey.org/~dugsong/dsniff

http://sectools.org

http://www.solarwinds.com

http://www.wildpackets.com

http://www.netinfo.tsarfin.com

http://www.colasoft.com

http://www.radmin.com

http://www.ipswitch.com/products/ws_ping

http://www.ipswitchft.com

http://www.filezilla-project.org

http://www.smartftp.com

http://www.cuteftp.com

http://www.coreftp.com

http://www.WinSCP.net

http://www.fireftp.mozdev.org

http://www.PuTTY.org

http://www.openssh.com

http://www.bitvise.com/download-area

http://www.bitvise.com/download-area

Vendor	Tool
Wireless Monitoring	
Wireshark	Wireshark
Wildpackets	Wildpackets
Tamo Soft	CommView for WiFi
Presentation Tools	
Adobe	Id, Ai, Ps, Fl, Dw, Acrobat
Microsoft	Word, Publisher, Powerpoint, Excel, Visio
Corel	Paint Shop Pro
Color Wheel	Color Picker
Techsmith	Snagit Screen Capture
Typography	
Int'l Typography Corp	Fonts
I Love Typography	Font Website
1001 Free Fonts	Free Fonts
Acid Fonts	Free Fonts
Fonts	Commercials Fonts
Urban Fonts	Free Dingbats
Blog Software	
Wordpress	Wordpress Blog Software
b2evolution	Multiblog Engine
Blogger	Blogger Software
Theme Dreamer	Dreamweaver Wordpress Extension
Video Tools	
uTIPu	uTIPu Screen Recorder
Techsmith	Camtasia Screen Recorder
Applian Technologies	Video Conversion Tool
Xilisoft	Video Conversion Tool
Smart Soft	Video Conversion Tool
Online Media Tech	Video Conversion Tool
Flip Cam	Video Camera

Location

http://www.wireshark.org

http://www.wildpackets.com/products/omnipeek/wireless

http://www.tamos.com/products/commwifi

http://www.adobe.com

http://www.microsoft.com

http://www.corel.com

http://www.pkworld.de/software/pkcolorpicker.htm

http://www.techsmith.com

http://itcfonts.com

http://ilovetypography.com

http://1001freefonts.com

http://acidfonts.com

http://fonts.com

http://urbanfonts.com

http://www.wordpress.org

http://www.b2evolution.net

http://www.blogger.com

http://www.themedreamer.com

http://www.uTIPu.com

http://www.techsmith.com

http://replay-converter.com

http://xilisoft.com

http://flvtoaviconverter.com

http://avs4you.com

http://theflip.com

Vendor	Tool
Voice Software	
Adobe	Audition
Twisted Wave	Twisted Wave
Audacity	Audacity
Voice Tools	
Harlan Hogan	Porta Booth
Marshall Electronics	MXL 990 USB Stereo Conderser Mic
Blue Microphones	Blue Snowball USB Microphone
CEntrance	MicPortPro USB Mic Preamp

Location

http://adobe.com

http://twistedwave.com

http://audacity.sourceforge.net/

http://harlanhogan.com/portaboothArticle.shtml

http://www.mxlmics.com, http://musiciansfriend.com

http://bluemic.com, http://musiciansfriend.com

http://www.centrance.com/products/mp

12
High Performance Teams

High performance products and market leaders come from high performance organizations. I have been successful at Technical Marketing on my own. I have also built highly successful teams. It can be done either way, however, there is more power in numbers. How do you create the environment that will encourage employees to work together successfully to achieve successful critical mass?

An essential element of getting everyone on the same page is securing an ownership of responsibility from each employee down to the lowest level of the organization - creating an on-going program to develop and maintain their ownership, their personal responsibility for producing results. Also recognizing when you've got a bench warmer vs. a star player early on will be crucial to your success. You can't have any bench warmers on your team, it must be all star players. Recognizing technical competence vs. technical incompetence early on can make or break the team.

Some would argue that developing buy-in is merely a step in effective delegation or delegating to the proper level. For delegation

> " It is the ownership of responsibility to achieve their own results, that unleashes initiative and provides satisfaction and achievement. "

to be effective, however, a framework of support must be in place - a framework for reinforcing consistent practices throughout the organization that rewards people for producing outstanding results.

A business operating on a foundation of objectives which every employee is committed to achieving, provides the focus for performance. Clearly defining and obtaining employee commitment to achieving the objectives directs employees toward meeting or exceeding objectives. When employees are performing together on that level, it generates high morale and high morale drives exceptional performance. High morale is the single most powerful motivation - even above financial - that drives exceptional performance on a team.

An effective foundation of common objectives can be broken down into eight essential elements.

Vision

A vision statement expresses a measurable ideal for the organization to reach at some point in the future - where you are going. These ideals become the passion for the team and the employees. Try to identify objectives the employees would boast about achieving, ones that inspire "*esprit de corps*." A classic vision statement once came from JFK when he said "We will put a man on the moon in a decade." All good Technical Marketeers have vision and know how to articulate it.

Values

Values are the basic virtues, the ethical backbone of the company's daily actions, for everyone working in the organization. Values are real issues employees can feel, believe, see, understand, and commit to. Grab onto some virtues that employees can hold true to, and that are necessary for the team to operate effectively. Examples include honesty, integrity, self-discipline, high customer satisfaction, and high productivity levels.

Mission

A statement expressing the value proposition the business delivers to its clients in the simplest, basic form. It is the answer to "What is in it for the customer?"

Motivation

It is easy to motivate yourself, not easy to motivate your team. It is the *"ownership of responsibility to achieve their own results"* that unleashes initiative and provides satisfaction and achievement. This is the fireball that produces outstanding results for the organization. It must be set forth initially and maintained regularly to compete and succeed - as long as you have star players.

Business Plan

A document containing the detailed plans for executing the mission and the vision: the management team, organization structure, financial plans, competition assessment, market differentiation, and product or service descriptions.

Marketing Plan

A document containing the detailed plans for creating the demand and producing sales for the products and services offered by the organization. Be sure to include Technical Marketing plans.

Goals and Responsibilities

A series of documents that establish the goals and responsibilities for each person in the organization, the results and deliverables expected, a timeframe for delivery, guidelines they are to work within, and resources they are to utilize.

These must be reviewed on a weekly basis because of the nature of change in business. To be successful, the organization must reinforce each employee in knowing their own goals, responsibilities and deliverables and working to achieve them. The organization must reward them for exceptional performance. Like the business plan, goals and responsibilities are an active plan of attack.

Many organizations have adopted the S.M.A.R.T. goals system and there is online tracking software that measures employees performance based on this. Goals must be Specific, Measurable, Attainable, Realistic and Timely. Using S.M.A.R.T. makes goal achievement powerful.

Organization Plan

A plan that allows the staff to understand the steps and processes in the organization - displaying what happens before and after each employees involvement.

As with the other elements of the foundation, the organization plan must be known and understood by every employee in the organization to ensure effective execution of procedures in their own area.

Operating Protocol

Identifies the actions required to ensure the foundation is continuously communicated, reviewed and updated as needed. Operating protocol assures the focus of the organization is expressed frequently throughout. This helps to ensure that each employee knows and understands the foundation.

13
Certifications

Certifications have great educational value. Read the books, don't take the tests. I have wasted a lot of time and money on certifications, and I have a lot of them. At the end of the day, certifications don't matter to a Technical Marketeer. The IE's that I know only have big ego's and they can't be used for Technical Marketing. Unfortunately, certifications have evolved into a hoaky way to get people brainwashed on a certain company's product at the consumer's expense of time and money. For Technical Marketeer's, it is a waste to channel your time and money in that direction. It would be like running the opposite direction in a race. Certifications are good for support and IT professionals, yet bad for Technical Marketeers.

If a product is so complex that it takes umpteen levels of certification, thousands of dollars to support it, your own money to pay for the certification, your own money to buy the equipment and books to educate yourself, along with a big ego and the correct alignment of the stars and planets to keep it running, it isn't worthy of today's fast and lean datacenters, nor your time. If you see a company that has a product with a big certification program, I advise you to run the other direction.

Users are smart enough to figure things out. If a product isn't easy enough to use, the marketplace with reject it. Products are easier to use now, that it only takes 3 clicks to get a product up and running. Get the product installed, running, document it and move on. If you don't, or if you can't, the war will be fought and won by the time you poke your head out of the lab.

Certifications actually distract raw talent away from the development of brilliant ideas for companies. Certifications result in a reduction in GDP and contribute to a decline in the economy. This is an unproductive activity, so how many times and how many ways can I say, "avoid them".

Avoid certifications because, let's face it, unlike a college education, certifications do expire. The companies that issue them can and will take them away from you, the tests become devalued as they are sold on the internet anyway, they are temporary, they don't carry any weight on your resume and they don't have the accreditation as that of real universities and colleges.

The Exit Strategy

There is always an exit strategy. Typically, it is either Initial Public Offering (IPO) or Takeover. Initial Public Offerings generate a lot of cash for the organization and have historically made founders rich. But the revenue in the company and the market has to be there to support the stock price and continued growth. Otherwise, IPO's end up where they started, just a windfall, the founder's cash out and the technology fades. Takeovers still generate cash for the founders and have a higher hope of the technology surviving and taking on new life, and because it is less dependent on capital markets this has been the exit strategy as of late.

Organization Profile questions are important, so know this language yourself. Because in a few quick sentences takeover artists, mergers and acquisitions kings can size you up faster than you can blink just by asking some of these key questions. All of which give credence to the performance of an organization or lack thereof, competitively.

Whether this information is used in competitive analysis or not is up to individual preference. You most often find this trudgery being dished out to industry analysts so that they can accurately spin your company to the street or industry.

The answers to the questions on the Organization Profile sheet will allow you to quickly determine how your organization compares to a competitors. Do you have high ground in the marketplace, are you always getting to the customer first, are you winning most of the battles? Is there significant growth in the market to take advantage of or is it time for an acquisition or to be acquired?

15

Final Remarks

For Professionals

If you are already working in technical marketing, hone your skills. If you are thinking of getting into technical marketing, be it known unto you that you must really enjoy hard work, and must have a natural passion for this type of work. Technical Marketeers love sitting in the lab and plugging in equipment. They love running tests and gathering results. They love looking through test results and competitive data to find a story or a competitive delta that didn't exist before. Some technical marketers take it to the next level and engage in pure strategy and charting strategic direction for the companies they work for. Some technical marketeers find a healthy career in product management and/or product marketing. There are many levels of technical marketing to engage your skills and there is always a lot of work to be done, so find the passion and help create value in bridging the gap between engineering and sales.

For Students

I couldn't decide what to do with my life when I entered college, but I did have what I would call today, a gift. I entered my first year of college as a Computer Science major. The program got really intense during my Junior year, and I didn't like it. Book learning wasn't for me. We were getting too deep into bits, bytes, gates, ands, nands, ors, microprocessors and such. There were long hours in the labs, writing programs in Pascal to model databases and queueing theory. We built operating systems in assembler language using punch cards on old IBM typewriter machines that were pre-mainframe era. To run the program, you had to stack and feed your cards into the card reader. We carried our programs around in shoeboxes full of punchcards. What was bitterly comical is when the box tiped over and the cards flew out, having to put them back in order by hand manually. In an advanced class I remember using a monochrome (black and white) monitor that ran at 300 baud back to the campus mainframe and I thought this was smokin because it was better than punchcards. The hardest class was #151 and I felt like I needed a bottle of it just to get through. I contemplated hard at changing my major to finance, and even shopped around for different colleges in different parts of the state. I was about to do it, when an HR manager at my father's company told me not to. I don't know why I listened to that guy, because some people would have thought he was a jerk. I will say he was very direct, honest and correct. Nevertheless, as he said, "use your gift, get your BS in Computer Science, then get your Masters and you can write your ticket anywhere". He was right. I would have been a terrible lawyer, even though I can read and understand legal briefs and I file my own patents, copyrights and trademarks today. I would have been a terrible doctor, at least I would not have enjoyed it that much. I actually ended up doing what I do best, and I love what I do. All I can say, and it was said to me once. "Do what you like, and like what you do. The rest will fall into place".

Index

81

Craig Thomas Ellrod has more than twenty years of experience in the computer industry and holds a Bachelor of Science in Computer Science from California State University, Chico, and a Masters in Business Administration from Pepperdine University. He has held many positions in the computer industry including software programmer, technical support, field and corporate system engineering, technical and product marketing, product management and sales. He has worked for companies such as Celerity Computing, Emulex, Pinnacle Storage, Sync Research, Cisco Systems, Extreme Networks, Citrix Systems and smaller startup ventures. He has authored several patent applications, patent designs and received an innovation award while at Extreme Networks.

Craig is the founder of the Technical Marketing Group based in the heart of Silicon Valley, and runs a Technical Marketing Blog and online forum. He is a legacy to the computer industry, as his father was an early pioneer in the computer industry.

His father held several positions with companies such as IBM, Control Data, NCR and several other smaller companies and startup ventures. For the majority of his career, his father headed up advanced systems laboratories, held positions in product development, manufacturing, and corporate management, finishing his career as a CEO. In the late 50's, he developed an optimized search algorithm for data storage technology using matrix algebra to express the migration of storage patterns. At the time, the computer used to perform this work occupied an entire room in the basement of the EE Dept at the university, as computers ran on vacuum tubes, and the silicon chip microprocessor had not yet been developed.

Technical Marketing®

secrets revealed

Strategy and Tactics for Competitive Results

"The Silicon Valley Survival Handbook"

first edition

by
Craig Thomas Ellrod

Publisher:

Technical Marketing, Copyright © 2008 by StrateQuest, LLC. All
rights reserved. This publication contains information protected
by copyright and trademark laws. No part of this publication may
be used, photocopied or reproduced in any manner whatsoever
without prior written consent from StrateQuest, LLC. For
information, address to StrateQuest, LLC., 2068 Walsh Ave Suite
B2, Santa Clara, CA 95050 U.S.A. SAN: 8 5 7 – 6 8 7 4.

ISBN-13: 978-0-9822570-0-5
ISBN-10: 0-9822570-0-7

Blog: http://technicalmarketing.org
Web: http://www.technicalmarketinggroup.com

ISBN 978-0-9822570-0-5

56995

9 780982 257005